My First Science Words

Earth Words

A LITTLE HONEY BOOK

Taylor Farley

Crabtree Publishing
crabtreebooks.com

desert

(DEZ-urt)

mountain

(MOUN-tuhn)

volcano

(vol-KAY-noh)

canyon

(KAN-yuhn)

ocean

(OH-shuhn)

island

(EYE-luhnd)

river

(RIV-ur)

forest

(FOR-ist)

lake

(LAYK)

prairie

(PRAIR-ee)

Glossary

canyon (KAN-yuhn): A canyon is a deep, narrow valley with steep sides.

desert (DEZ-urt): A desert is a dry area where there is little or no rain.

forest (FOR-ist): A forest is a large area covered with trees and plants.

island (EYE-luhnd): An island is a piece of land surrounded by water.

lake (LAYK): A lake is a large body of water surrounded by land.

mountain (MOUN-tuhn): A mountain is a very high piece of land.

ocean (OH-shuhn): An ocean is a body of salt water.

prairie (PRAIR-ee): A prairie is a large area of flat grasslands with no trees.

river (RIV-ur): A river is a large stream of fresh water that fl ws into a lake or ocean.

volcano (vol-KAY-noh): A volcano is a mountain that has openings from which lava, ash, and gas sometimes erupt.

School-to-Home Support for Caregivers and Teachers

Crabtree Seedlings books help children grow by letting them practice reading. Here are a few guiding questions to help the reader build his or her comprehension skills. Possible answers are included.

Before Reading

- **What do I think this book is about?** I think this book is about different places on Earth. I see a forest, volcano, island, river, and mountain on the cover.
- **What do I want to learn about this topic?** I want to learn about more places on Earth. What do they look like?

During Reading

- **I wonder why...** I wonder why the desert has huge rocks with different shapes. Some have flat tops and sides. Others are pointy.
- **What have I learned so far?** I have learned the words desert, canyon, ocean, lake, and prairie. The canyon looks very deep. The prairie looks very flat.

After Reading

- **What details did I learn about this topic?** I learned that a river can have a bendy shape. It is different from the straight river in my community.
- **Write down unfamiliar words and ask questions to help understand their meaning.** I see the word ***canyon*** on page 8 and the word ***prairie*** on page 20. The other vocabulary words are found on pages 22 and 23.

Crabtree Publishing

crabtreebooks.com 800-387-7650

In Canada: We acknowledge the financial support of the Government of Canada through the Canada Book Fund for our publishing activities.

Hardcover 978-1-4271-3043-3
Paperback 978-1-4271-3048-8

Printed in the U.S.A.
042025/CP20250402

Published in Canada
Crabtree Publishing
616 Welland Avenue
St. Catharines, Ontario
L2M 5V6

Published in the United States
Crabtree Publishing
347 Fifth Avenue
Suite 1402-145
New York, NY 10016

Written by: Taylor Farley

Photo Credits: All illustrations by aekikuis; page 3 © Shutterstock.com /Angelo Ferraris; cover and page 5 © Shutterstock.com /Taras Kushnir; cover and page 7 © Shutterstock.com / Fotos593; page 9 © Shutterstock.com /Galyna Andrushko, page 11 © Shutterstock.com / TouchingPixel; page 13 © Shutterstock.com /Vibrant Image Studio; page 17 © Shutterstock.com /Olga Danylenko, page 15 © Ken Backer | Dreamstime.com, page 19 ©istock.com/ skiserge1, page 21 ©istock.com/ SharonDay

Library and Archives Canada Cataloguing in Publication
Available at the Library and Archives Canada

Library of Congress Cataloging-in-Publication Data
Available at the Library of Congress